*World of science*

# MAGNETISM AND ELECTRICITY

KU-174-680

**BAY BOOKS LONDON & SYDNEY**

1980 Published by Bay Books
157-167 Bayswater Road, Rushcutters
Bay NSW 2011 Australia
© 1980 Bay Books
National Library of Australia
Card Number and ISBN 0 85835 273 7
Design: Sackville Design Group
Printed by Tien Wah Press, Singapore.

# MAGNETISM AND ELECTRICITY

Magnetism and electricity are not completely understood, but in one form or another these two forces affect almost everything we do. Almost every piece of modern machinery relies on electricity in some way; we use batteries and generators in motor cars, electric power to light and heat our homes and to operate ovens, refrigerators, washing machines and so on.

## MAGNETISM: The lodestone and the compass

The ancient Greeks became acquainted with magnetism a long time ago when they discovered the amazing properties of a certain type of iron ore. It was found that when a small sliver of this metal was suspended from a string it always pointed in more or less the same direction. They called it *lodestone* and the direction it pointed was

This early compass was used by Italian mariners in 1580 to navigate at sea. The vellum card is mounted inside a case made of real ivory.

This home-made compass can be easily made using a few household items. All you need is a piece of card, 3 strong needles and some cork. Construct the compass as shown in the diagram, making sure that the north poles of the 2 horizontal needles are both pointing in the same direction.

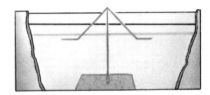

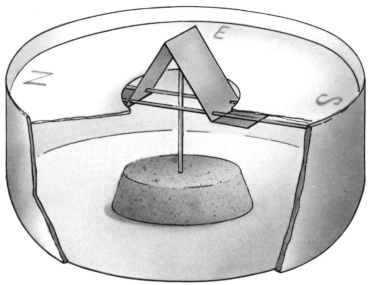

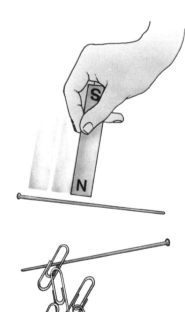

You can magnetise an iron nail by stroking it with a magnet. The magnetised nail will attract and even pick up small metal paper clips.

always north and south. This became the first type of compass and was used by many of the early navigators. Experiments with lodestone led to the discovery that pieces of iron could be magnetised by stroking with lodestone and that a magnetised iron needle could be pushed through a piece of cork and floated in a wooden or copper bowl of water so that it would rotate more freely than if it was suspended on a string. The compass used at the time of Columbus was of this type.

## Magnetising metal

The simple pocket compass you can buy today is similar to the original one. It is accurate enough for bushwalking and general direction finding. Its needle consists of a thin strip of metal which has been artificially magnetised. Many metal alloys containing iron are capable of being magnetised. Iron and steel make magnets as do mixtures of nickel, cobalt and iron. Some metals such as soft iron are easy to magnetise though they lose their magnetism just as easily. They are used to make magnets that can be turned on and off, such as the electromagnet. Hard steel and some alloys are difficult to magnetise but may hold their magnetism for a very long time. With a bar or horseshoe magnet it is easy to make another one. It is only necessary to stroke a magnet along a small piece of iron or steel such as a knitting needle. The unmagnetised metal is stroked a number of times in the same direction, always using the same end or *pole* of the magnet. When

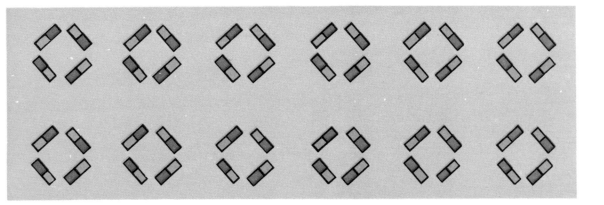

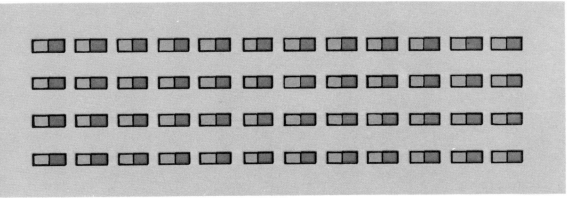

the piece of metal is magnetised it gains the property of attracting iron, steel and their alloys. A magnet attracts a steel paper clip because the clip is magnetised by the magnet.

Iron, steel, cobalt and nickel are known as *ferromagnetic* metals. Metals, like other materials, are made of millions of atoms, if they are pure elements like iron, or of molecules, if they are compounds like steel. Within the atoms, electrons revolve around the nucleus and as they spin a *magnetic field* is created. This field is like that of a miniature magnet. When the ferromagnetic metals are unmagnetised, these mini-magnets are so arranged that they point in all directions, cancelling each other out. But when the metal is magnetised, many of these mini-magnets line up and point in the same direction, that is, with their north poles pointing one way and their south poles pointing the other way. The combined effect is to create one strong magnet.

If you drop a magnet, hit it with a hammer, or heat it in a fire, it becomes demagnetised again. All the mini-magnets again point in different directions and once again cancel each other out.

Top: The north and south poles attract each other when a piece of iron or ferromagnetic metal is unmagnetised. Thus they cancel out the magnetism.

Above: However, when magnetised, the mini-magnets line up with the north poles pointing one way and the south poles another way to create an extremely powerful magnet.

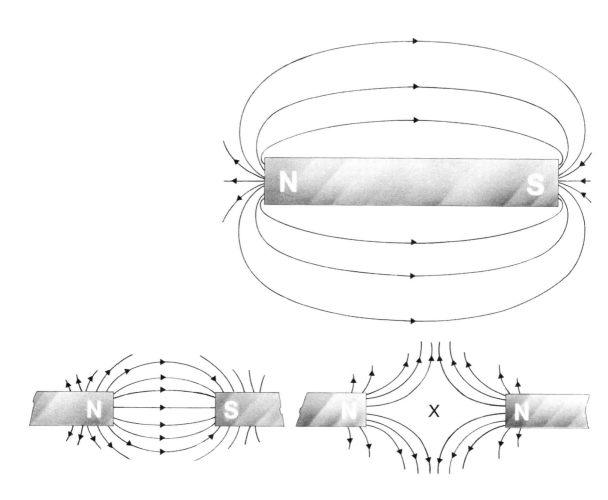

Like poles repel each other whereas unlike poles attract. The magnetic force lines are shown in these diagrams. At the point X, no magnetic force can exist.

# Magnetic poles and magnetic field

A magnetised bar has two magnetic poles, one at each end. They are called the *north* and *south* poles. A fundamental rule of magnetism is that opposite poles attract each other and similar poles repel each other. If you take two bar-magnets and place the ends next to each other, the north of one will attract the south and repel the north of the other.

The end of a compass needle that points north is always marked, so that it is possible immediately to tell which is north.

All magnets have a *magnetic field,* which is the area of influence of the magnet. Normally, the magnetic field is invisible but it can be revealed by an interesting experiment with iron filings. To do this, take some iron filings (which can be obtained simply by filing a piece of iron with a file), a magnet and a piece of cardboard. Place the cardboard on the magnet and sprinkle some iron filings on it. Give it a gentle tap or two and the filings will jump into a pattern representing the magnetic *lines of force,* or *magnetic field,* of the magnet.

It is interesting to do this with two bar-magnets, standing side by side on end; first with similar poles together, then opposite poles. Move the magnets under the paper and see what happens to the iron filings in each case. Try the same experiment with a horseshoe magnet.

When you use a bar-magnet to demonstrate lines of force you will be able to see quite clearly how the greatest concentration of magnetic force is at the end of the magnet and the lines of force are curved, running from one pole to another. The earth with its magnetic poles has a similar type of magnetic field.

A simple experiment with iron filings can demonstrate the magnetic field of a horseshoe magnet. The unlike (north and south) poles attract each other and the iron filings form a pattern to represent the magnetic lines.

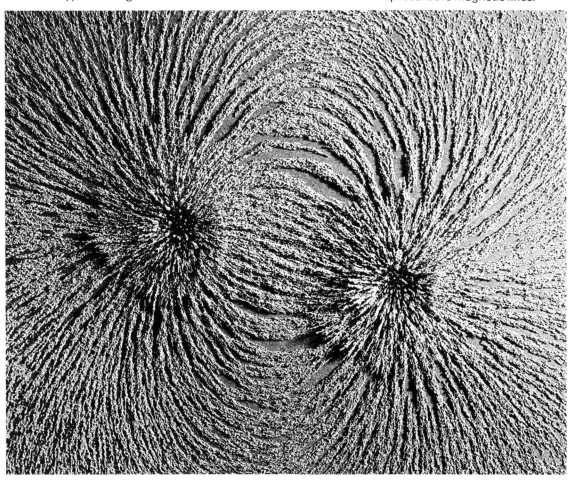

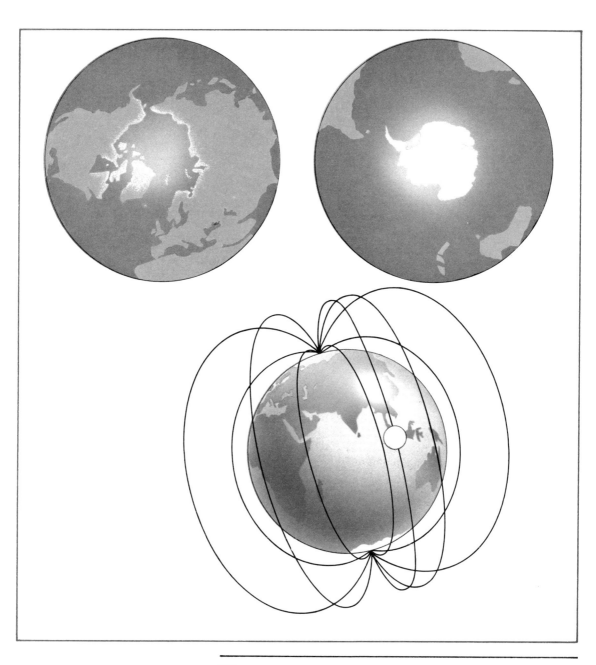

The earth is an enormous magnet with its own magnetic field and magnetic north and south poles which are situated in Canada and the Antarctic respectively.

# The earth's magnetism

The earth itself is a huge magnet. Like a bar magnet, it has two *magnetic poles*. The *north magnetic pole* is about 1,600 km from the geographic North Pole. The *south magnetic pole* is about 2,400 km from the geographic South Pole. The angle between the directions of magnetic north and true north at any place is called the *magnetic declination* or *variation*. This angle varies from place to place and, because the magnetic poles change position, from year to year.

The earth's magnetism varies from place to place because of the differing nature of rocks and it is therefore possible to locate metals by the movement of a magnetic needle in an instrument called a magnetometer.

From time to time, great bursts of activity called *sunspots* take place on the surface of the sun and cause *solar winds* to sweep the earth. These solar winds are storms of electrically charged particles which penetrate and affect the magnetic field of the earth. When this happens, compass needles swing in all directions, making navigation with the magnetic compass extremely difficult. Radio communications are also affected, with large amounts of static.

Above: All planes have a gyrocompass in the control panel for navigation purposes (see centre top left).

# The ship's compass

The ship's compass, or mariner's compass, shows a card instead of a needle. The card is marked to show points of the compass around the 360 degrees of a circle. On the underside of the card are strips of magnetised metal. The

Below: This cross-section through a mariner's compass clearly shows the card on which all the compass points are marked, floating in the mixture of alcohol and water that fills the compass bowl. When a ship rolls in stormy seas, the compass still gives accurate readings.

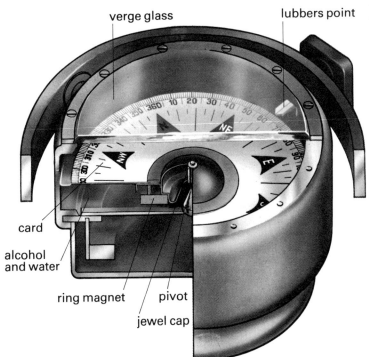

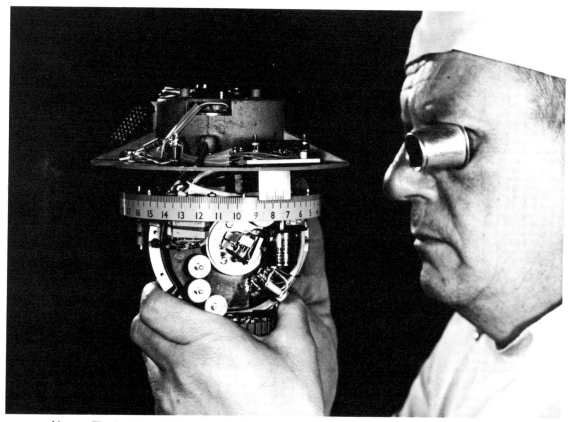

Above: The inside working mechanism of the Sperry CLII rotorace gyrocompass, used in military and commercial planes.

Below: A pendulous gyrocompass will always adjust its axis towards the North Pole.

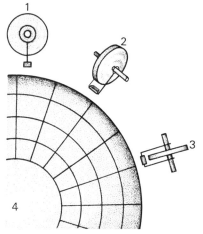

card floats in a bowl filled with a mixture of alcohol and water and is free to swing in any direction. The floating action also means that, as the ship rolls, the compass card remains more or less in a steady position. The bowl in which the card floats is mounted in such a way as to prevent the compass from swinging about too wildly in rough weather. The mountings are called *gimbals*.

There are some problems connected with using a magnetic compass in ships and planes. The first of these is the masses of metal in the craft itself. These tend to cause the needle to 'deviate' or turn at an angle from its true direction. Although it is possible to compensate for this in various ways, there is also disturbance caused by deposits of iron close to shore and by the presence of other ships. To overcome this, most ships and planes now use a *gyrocompass*, which is not affected by these problems.

The gyrocompass is based on the use of a *gyroscope*, which is a heavy disc spinning rapidly on an axis suspended in gimbals so that as the plane or ship moves or rocks in different directions the axis of the gyrocompass swings to point in a constant direction.

# ELECTROMAGNETISM

*Electromagnetism* is magnetism produced with the help of an electric current. The word also means the study of the relationship between electricity and magnetism. The scientific study of electromagnetism began early in the last century when the Danish scientist Hans Christian Oersted noticed that a magnetic needle was deflected from its usual direction when an electric current flowed through a nearby wire. Because magnetic poles attract or repel each other, he concluded that an electric current must produce a magnetic field. His work in this field helped other pioneers such as André Ampère and Michael Faraday to make important new discoveries.

## Oersted's experiment

What Oersted observed can be seen by a simple experiment. The materials required are a single strand of wire passing vertically through a small hole in a piece of cardboard and connected to a battery, and some iron filings. When you switch on the current and sprinkle the iron filings on the cardboard, the filings will form a series of concentric circles around the wire. Gently tapping the cardboard helps the experiment along. The iron filings are aligning themselves along the lines of force in the same manner as with a magnet. As magnetism in a material is determined by the movement of electrons and an electric

Above: In 1820, the Danish scientist Hans Oersted discovered the relationship between magnetism and and electricity. When an electric current was directed through a wire, a nearby magnetic needle changed direction.

Left: Oersted passed an electric current through a piece of card sprinkled with iron filings. The filings spread out to form concentric circles, aligning along the magnetic lines of force.

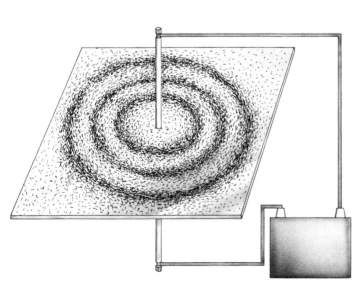

This apparatus was used by Oersted in his 1820 compass experiment when he discovered that an electric current passing through the overhead wire deflected the compass needle below, thus establishing the relationship existing between magnetism and electricity.

Below: In a solenoid, a coil of insulated wire is wound around a central iron tube. The strength of the magnetic field produced when a current is passed through the solenoid is determined by the number of turns in the coil of insulated wire.

current is a flow of electrons, the current gives rise to a magnetic field. Of course, the effect shown in this experiment depends on the power of the electric current and the position of the wire in relation to a field of iron filings or to a magnet.

## The solenoid

A strong magnetic field can be produced by passing a current through a *solenoid*. This is a tube around which is wound a coil of *insulated* wire. *Insulated* wire is coated with varnish or plastic. Each turn of the coil produces a magnetic field and the fields combine to make one powerful field. The tube of the solenoid behaves like a bar magnet with its north and south poles. The strength of the magnetic field can be increased by placing a bar of soft iron inside the tube. Naturally, the iron also becomes

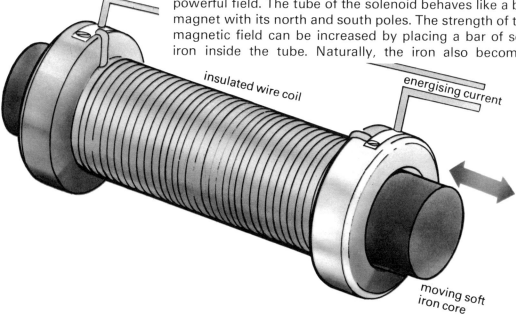

magnetised. The tube and the iron bar are the core and together with the coil or wire make an electromagnet. The magnet works when the current is on, for the soft iron becomes magnetised very quickly. On the other hand, when the current is switched off, the soft iron loses most, though not quite all, of its magnetism just as quickly. This is very convenient for it means that when the magnet is not required it may be switched off and will not attract metal.

# The electromagnet

The first electromagnet was built by an Englishman, William Sturgeon, in 1825. He bent a soft iron bar, insulated it by coating it with varnish, and wound 18 turns of copper wire around it. When a current from an electric cell was passed through the coil, the seven ounce bar (about 200 g) was able to lift another iron bar weighing

An electromagnetic crane can be used to lift incredibly heavy objects such as this large slab of solid steel.

seven pounds (about 3178 g). An improvement to this simple device was made in 1831 in the United States by Joseph Henry, who found it was better to insulate the wire rather than the bar. Inside an electric motor you will see the same idea at work in the insulated wire wrapped around bare metal.

Some of the largest electromagnets are extremely powerful, but they produce great heat. To cool them down, pipes carrying water are embedded in the coil that carries the electric current. In some types, the coil is made of hollow wires through which water can flow. An electromagnet at the National Magnet Laboratory, Massachussets Institute of Technology, requires about 9000 litres of cooling water every minute.

# ELECTRICITY

Electricity is a form of energy that is normally invisible. We know about it by what it does rather than by what it looks like. The most obvious effect of electricity in nature is observed in a lightning flash. Electricity travels through some substances, like copper and iron; these are called *conductors*. However electricity will not pass through some materials like rubber and glass, and these are called *non-conductors* or *insulators*.

## Static electricity

Another effect, observed as long ago as 600 BC by the Greek philosopher Thales, is that a piece of amber that has been rubbed will attract tiny objects towards it. A curious connection between electricity and magnetism is that the word 'electricity' comes from the Greek 'electron', which means 'amber', and the amber used by Thales behaved very much like a magnet. In the 1700s the

Right: Try out this simple experiment for yourself to demonstrate static electricity. Just rub an amber or glass rod with a piece of silk and you will see that it attracts pieces of paper.

Opposite: These men are busy repairing damaged insulators on a transmission tower. Glass or porcelain insulators are used as rigid mountings to support power cables.

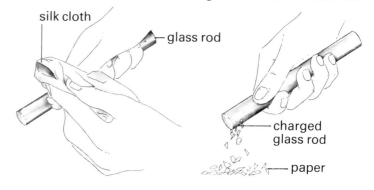

American politician and scientist, Benjamin Franklin, flew a kite in a thunderstorm to show that lightning is caused by electricity. The difference between the electricity of the lightning flash and that of rubbed amber is that the electricity in the amber is still, or *static*, electricity.

The knowledge of electrification by rubbing or *friction* led to the invention of machines such as the Wimshurst machine. Invented in 1882, it consists of two glass discs mounted side by side on a single axle and rotating in opposite directions. Around the edge of each disc are a number of flat metal foil contacts. Contacts on opposite sides of each disc are connected to a fixed metal conductor by soft wire brushes touching the foils. The machine is turned by hand and as the discs rotate friction produces positive and negative charges.

The Wimshurst machine, invented in 1882, generated static electricity by means of two rotating glass discs which were mounted on a single axis.

# Producing electricity with chemicals and magnets

The first electric battery was invented by the Italian Alessandro Volta, who built a simple battery which was called a *voltaic pile* in 1800. The *volt*, a standard measure of electrical force, comes from his name. His battery worked on the principle that, if a plate of copper and a plate of zinc are placed in a bath of sulphuric acid, an electric current will be created and will flow from the copper to the zinc. This is the simple voltaic cell, the earliest form of electric battery.

Many scientists were exploring the properties of electricity in the first half of the nineteenth century. One of these was Faraday, who discovered the important

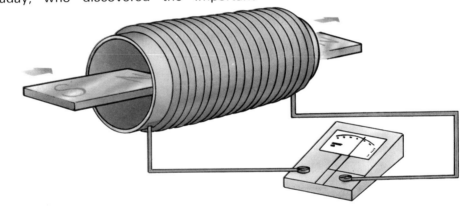

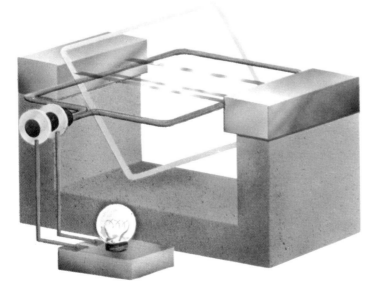

A bar-magnet moving within a coil (above) produces a momentary electric current which can be detected by a sensitive ammeter. The simple generator (left) consists of a wire loop which can rotate in a magnetic field

principle of *electromagnetic induction,* on which the *generation* and use of electricity is based.

If a wire coil is rotated between the poles of a magnet, an electric current will be produced in the rotating coil. This is the principle of the electric generator. The movement of a wire carrying electricity, the movement of the electric current in the wire and the movement of a magnetic field around the wire are all basic principles of the construction of electromagnets, electric motors and generators. In all of them there is a coil of wire, a core, like a bar-magnet, and a spinning motion.

Right: This early electric telegraph key transmitted the dots and dashes of Morse Code.

Below: Wheatstone's punched tape perforator, transmitter and pen-tape receiver was an important step forward in developing the telegraph system.

# Electrical inventions

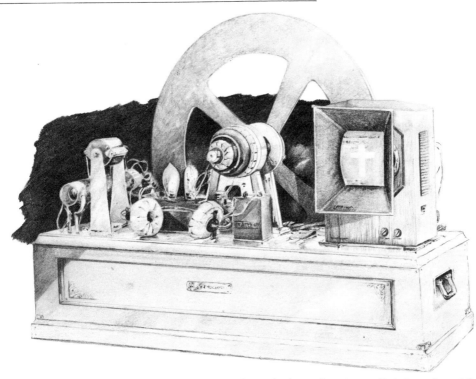

Above: Baird produced the first television system. This is the apparatus on which he transmitted a television picture, initially of a cross.

The discovery of these principles led to the foundation of the giant electrical industry. Electric power became available to everybody and electric motors replaced the steam engine as the mainstay of industry. Inventors, like Thomas Edison, found ways of using the effect of current passing through a wire and electric lights and electric heaters were developed. The first practical electric telegraph system was produced by Wheatstone and Cooke in 1837 and while they were working in England to improve telegraphy, Samuel Morse was doing the same in the United States. The first telegraph cable between Britain and France was laid across the English Channel in 1851 and a cable was laid between Britain and the United States in 1866.

Broadcasting and long-distance radio did not arrive until after World War I, when engineers realised they could make *amplifiers* to increase the strength of radio signals. A simple television apparatus was built in Britain by Baird in 1925 but modern television was delayed until the development of the *cathode ray tube* during World War II, when similar work was done on radar.

# ELECTRIC GENERATORS AND MOTORS

An electric generator converts or changes one form of energy, the motion of a coil or wire, into electrical energy. It does this by making use of Faraday's discovery that an electric current can be made to flow through a wire by moving the wire across a magnetic field. It is similar to an electromagnet, with the process reversed. Instead of electricity passing through the electromagnet, there is movement of the coil.

## Induction

All that is needed to produce electricity is a magnet and a piece of wire and energy to produce the motion. It does not matter if the wire or the magnet moves, so long as one moves in relation to the other. If the wire is in the form of a coil, then a voltage is *induced* or caused in each turn of the coil. The voltages add up so that a high voltage results. To make it more productive, several coils of wire are used.

## Generators

In most big generators, coils of wire are attached to a shaft and rotated within a magnetic field. The shaft is driven by a *turbine* operated by gas, steam or water power. The wire coils or *windings* are slotted into a metal drum called an *armature.* The magnetic field is supplied by

In this simple generator, an armature winding is attached to a central shaft and rotated within a magnetic field. The generated current is then collected by the slip rings and passed into the appliance by means of the brushes.

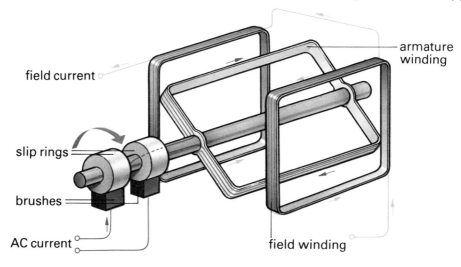

an electromagnet consisting of iron cores magnetised by coils of wire wrapped around them. The cores retain a small amount of magnetism at all times so that, when the generator starts up, the existing magnetism causes the first small voltage. Some of this is fed into the windings of the electromagnet and the magnetic field increases until the generator is operating at its full capacity.

This technician is carrying out maintenance work on an enormous electricity generator, the turbines of which are driven by steam.

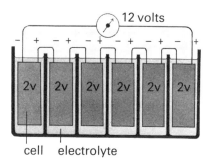

Above: Most car batteries consist of six 2 volt cells which are connected in series.

Below: This 'wet' car battery (lead-acid accumulator) has a lead cathode (negative terminal) and a lead dioxide anode (positive terminal).

As the field increases, more and more electrons begin to travel in the same direction within the wire. This *current* of electricity flows to where the end of the coil is connected to metal rings on the shaft. These rings, called slip rings, are spinning at the same speed as the shaft so it is necessary to have devices called brushes to take away the electricity. These are usually carbon rods, which are conductors of electricity. The brushes rest against the fast revolving rings, pressed down by springs to make sure they make good contact. The electricity passes through the carbon rods and is fed into a circuit.

## AC and DC

Electricity is generated whenever the armature windings cut across the lines of force of the magnetic field. Because the coil is rotating and the field is stationary, any one part of the coil is moving towards the field at one instant and

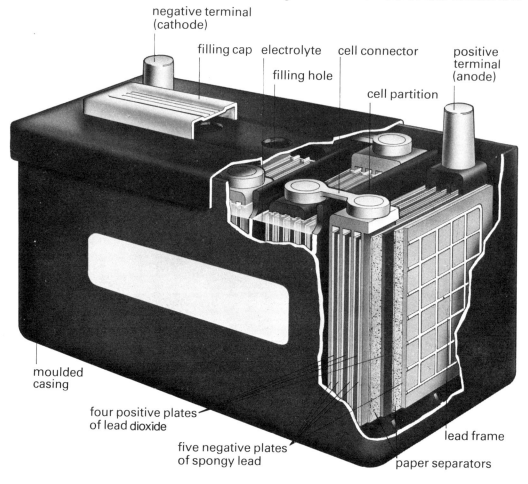

moving away from it the next. This means the electricity generated within the coil flows first in one direction and then in the opposite direction. In the generator described above, when the current is collected by the brushes touching the slip rings, *alternating current* results because the current in the coil is flowing first in one direction and then in the other.

Instead of slip rings, it is possible to have a collar divided into two or more segments at the end of the shaft so that before the current flowing through the winding changes its direction, another winding with current flowing in the same direction is connected to the brushes. All the current collected is flowing in the same direction and the result is *direct current*.

The electricity that is produced in the form of a current can be converted back to mechanical energy by means of an electric motor.

# Batteries

Batteries are used to provide electric current whenever it is inconvenient or impossible to draw power from the mains. We use batteries to operate the door-bell, to start the car, run flashlights and transistor radios, hearing aids and some cameras.

A battery is a set of *electrolytic* cells providing electricity as direct current. There are two kinds of cells.

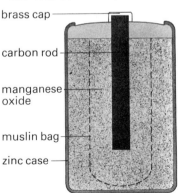

The dry cell battery (below) is used in torches and radios. Its compact size makes it convenient to use. The cross-section (below left) clearly shows the carbon rod connecting the two terminals. A paste of ammonium chloride is used – not a liquid as in the 'wet' cell battery version.

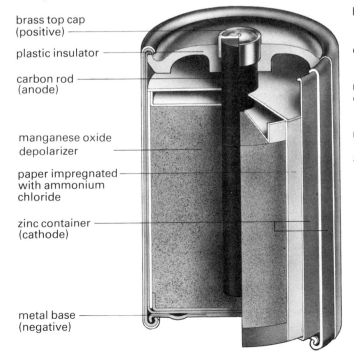

*Primary cells,* which produce their electricity by a chemical process which used up the chemicals, must be thrown away and replaced when they become exhausted. Flashlight or 'dry' batteries are an example. *Secondary cells* work on a reversible principle and can be recharged by connecting to another source of current such as a household power supply. They will last longer, though not indefinitely. A car battery or 'wet' battery is an example.

All cells have two *electrodes,* metal or carbon, in an *electrolyte,* which may be a fluid like a dilute acid or a paste as in most primary cells. When the two electrodes are placed in the electrolyte an excess of electrons is caused at one electrode and a deficiency at the other. The difference between the electrodes causes an electric current to flow. The tops of the electrodes such as we can see in a car battery are called the *terminals.* When the terminals are connected to a light bulb or electric motor the electric current causes the bulb to glow or the motor to turn.

The wet batteries used in a car are also called *storage cells,* because they are used to store electricity. These batteries can be *charged* from the power main or from the car's generator or alternator. When the charge has been used up, the battery can then be charged up again.

# Electric motors

An electric motor uses electromagnetism in the reverse of the way it is used in the generator. A motor consists of two main parts, one that is free to move and one that is fixed. The moving part is called the *rotor* or *armature,* the

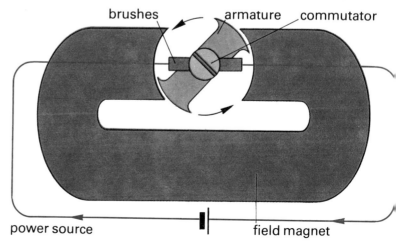

This simple electric motor consists of an armature that rotates inside a strong field magnet. The brushes make contact with the commutator and thus the direction of the current is periodically reversed.

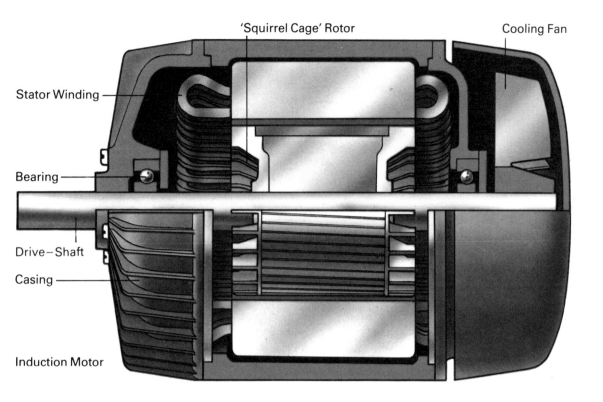

Induction Motor — 'Squirrel Cage' Rotor, Cooling Fan, Stator Winding, Bearing, Drive-Shaft, Casing

fixed part is called the *stator*. In a simple electric motor, the stator is a permanent magnet and the rotor consists of coils of wire wound on soft iron cores. As electricity is fed into the windings, an electromagnetic field with north and south poles is set up in the windings. These are attracted and repelled by the poles of the magnet so that the rotor spins around. By connecting up the end of the rotor shaft it is possible to use the motor to drive a machine.

Another alternating-current motor is the *induction motor*. Alternating current passes through fixed coils, producing a magnetic field which rotates. The effect is the

The induction motor, which is widely used in industry, has similar rotatory (rotor) and stationary (stator) windings, which together act like a transformer. The rotor and drive-shaft rotate by means of the magnetic field induced by the rotor.

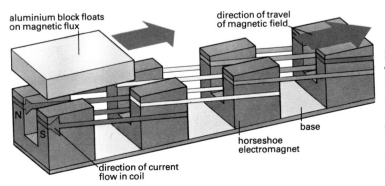

aluminium block floats on magnetic flux; direction of travel of magnetic field; base; horseshoe electromagnet; direction of current flow in coil

In the linear motor, the stator windings are arranged in a line (not a circle). Thus a moving magnetic field is produced which travels along the windings. This sort of motor is used in sliding doors and in baggage handling systems in large airports.

Large electric motors, such as this Otis variable speed AC drive motor, are used to power lifts in tall buildings.

same as if the stator were a rotating magnet. There is also a special type of alternating-current motor called the *synchronous motor.* It turns at the same speed as, or *in synchronism with,* the rotating field of the stator. Such motors are used in electric clocks and other devices where it is important to be able to maintain a constant speed of rotation.

Direct current motors have either permanent magnets or field coils with fixed magnetic poles. The electromagnetic coils on the armature are supplied with electric current through a device called a commutator, which is simply a ring on the shaft with separate contacts for each coil.

Another type of motor called the *linear induction motor* is sometimes used for electric trains. The principle of this motor involves using the engine and the track as if they were the rotor and stator of an electric motor. The current is carried on a live rail to a flat magnetic inductor. As the electromagnetic field is created, the inductor is pulled along by magnetic force in a horizontal direction instead of being rotated.

# CIRCUITS AND USES OF ELECTRICITY

A *circuit* is the pathway along which an electric current flows. It is usually wire or a thin strip of metal. An example of a simple circuit is that of a pocket flashlight. Electrons from the negative terminal of the battery pass through the filament of the bulb and return to the positive terminal. The circuit is opened or closed by means of a *switch*.

## Series and parallel circuits

Circuits can be in *series* or in *parallel*. To illustrate the difference, three bulbs can be connected to one battery. When they are in series the wire from the battery runs straight from one to the other and the circuit is controlled by a single switch. When the switch is open, no current flows and all the bulbs remain unlighted. Close the switch and all three light up. To connect the bulbs in parallel, separate circuits are made to run off the main circuit and switches can be placed at any point to open or close the circuit to any one or all of the bulbs. In a series circuit, if one bulb blows, all go out, but in a parallel circuit the failure of one bulb will not affect the others, so most lighting circuits are parallel circuits.

An electric cell can power both series and parallel circuits. The battery and the circuit (below) are represented diagrammatically in the form of symbols (below left). In the series circuit, all of the bulbs light up when the switch is closed. However, in the parallel circuit, when switch 1 is closed, only two bulbs light up. The bulb nearest to the battery will only light when switch 2 in the circuit is closed.

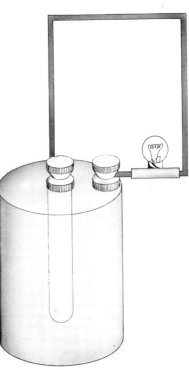

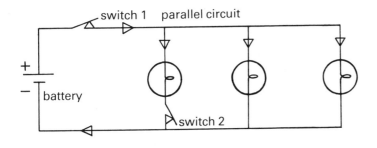

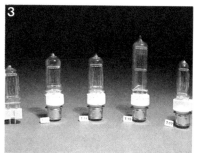

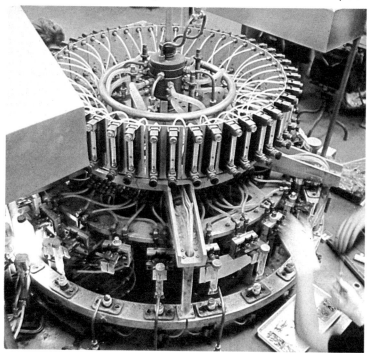

# Electric lights

Electricity can be converted to light in a number of ways. In the ordinary light globe electricity flows through a fine wire, heating it so much that it glows. The wire is called a *filament* and most modern filaments are made of coils of tungsten wire which, when heated, produce a strong white light. You can see the filament quite easily in a clear globe.

If it were heated white hot in the atmosphere, the tungsten filament would not last long, because the oxygen in the atmosphere would cause the tungsten wire to *oxidise* or burn. It would also lose part of its normal composition through the effect of heat and evaporation. To overcome these problems, the bulb has the air removed from it and nitrogen and argon gases are substituted. Most household bulbs use between 40 and 150 watts, but some searchlights and television studio tungsten lamps need as much as 30,000 watts.

*Discharge lamps* are built in the form of a long tube containing a gas or vapour. When electricity passes through gases at low pressures, energy is transferred to the gas atoms, causing them to radiate. The colour of the light depends on the gas that is used. Sodium vapour

Above: A selection of lamps. 1 and 2 Household lamps and bulbs. 3 Studio and theatre lamps. 4 Fire glow lamp. Right: These lamps are being assembled in a factory on an automatic production belt.

produces a bright yellow light and neon a strong red light.

*Fluorescent lamps* work by conducting electricity through mercury vapour, which causes ultraviolet radiation. Ultraviolet is invisible to our eyes but the inside of the tube is coated with a fluorescent powder. When the ultraviolet rays strike this powder, it glows with visible light.

*Carbon arc lamps* consist basically of two carbon rods fed with electricity. The electricity jumps from one rod to the other in the form of an *arc* or continuous spark and produces a very intense light. Arc lamps are used in some types of searchlights and motion picture projectors.

# The electric bell

The ordinary household bell works on the principle of electromagnetism. In the electric bell, one piece of iron called the *armature* is attracted towards another U-shaped piece of iron when the latter is magnetised. The armature is held against a contact by a spring. Pressing

How an electric bell works. When the button is pushed, the circuit is completed as the armature makes contact with the spring. Current flows from the battery, the electromagnet is activated and the clapper strikes the bell. Thus magnetism is used to make (or break) the circuit.

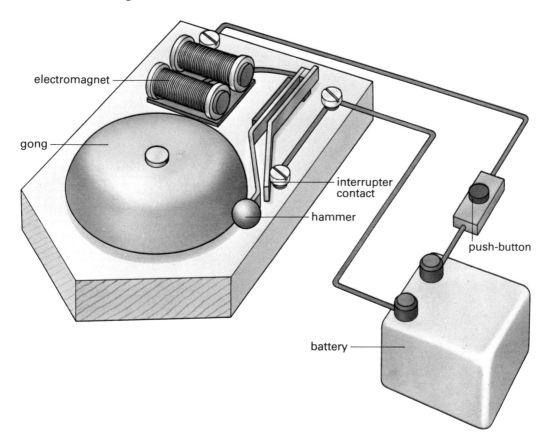

the bell-button starts the current flowing from a battery. The electromagnet is activated and the armature is pulled away from the contact and a hammer attached to the armature strikes a bell. But when the armature is pulled away from the contact, the electric current is temporarily cut off and the electromagnet loses its magnetism. The armature is then pulled back to the contact by the spring, the current flows again and the process repeats as long as the bell is pushed. Because it happens very quickly the result is the rapid ringing sound of the bell.

# Electroplating

This is the means of applying a very thin coating of metal, usually onto another metal. Many parts of the family car are plated, such as the chrome-plated bumper bar. By electroplating, a cheap metal, such as white brass or steel, can be given a thin coating of a more attractive metal, such as silver, or a protective coating of nickel or chrome. Most of the cutlery in use before stainless steel knives and

A brass key can be copper-plated by immersing it in a solution of copper sulphate together with a copper plate. When an electric current from a battery is passed through the solution, a film of copper is deposited on the key, and the copper plate is thus eaten away.

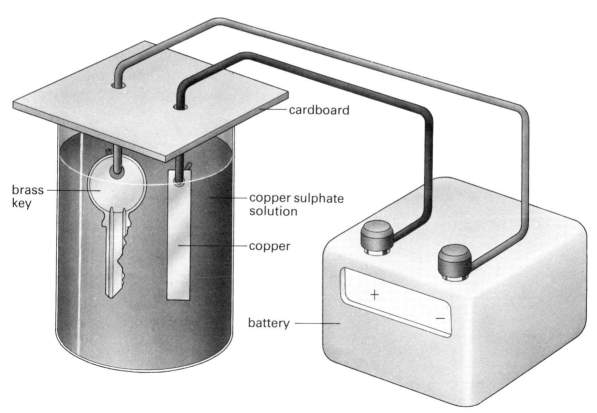

In this highly sophisticated automatic electroplating system, copper-plated printed circuit boards are produced.

forks became so popular had a coating as thin as one fifty-thousandth of an inch of silver. To make the silver last better, a coating of nickel is applied before the silver, hence the familiar letters E.P.N.S., which stand for electroplated nickel silver.

The reason why chromium is used so frequently, apart from its bright finish, is that chromium plating is hard and long-lasting. A development is 'hard chromium' plating in which a much thicker film is electroplated onto machinery parts to restore them. This technique is also used to harden the tips of tools.

The technique of electroplating depends on *electrolysis*. The article to be plated and a plate of the metal to be applied are placed in a bath containing a solution of the salt of the metal. A current of electricity is passed through the solution, using the part to be plated as the negative pole or *cathode* and the plating metal as the positive or *anode*. When the current is switched on, a film of metal is deposited onto the article while the metal plate is gradually eaten away. The first use of electroplating was the silver-plating of knives and forks about 100 years ago.

# The electron microscope

Some objects are too small to be studied through an ordinary microscope using light. In this case, an electron microscope is used.

In the electron microscope, a thin beam of electrons is focused onto the object to be examined by a magnetic coil, in much the same way as focusing light with a glass lens. Some of the electrons pass through the specimen to be magnified and others are scattered by the atoms in it. The electrons which pass through the specimen and those that are scattered are passed through another set of magnetic coils to form a magnified image on a fluorescent screen. On the screen the magnified image can be observed or photographed. The screen is somewhat like the screen of a television set, and those parts of the specimen will show up more brightly where fewer electrons have been scattered. The specimen must be very thin for electrons to pass through it. For thicker

Using a Cambridge *Stereoscan* electron microscope (above and right), you can not only magnify the object over 500,000 times using electron beams, but the microscope can also scan the object. Thus the operator can study the magnified image on a screen.

Opposite: A cross-sectional view through an electron microscope showing the ray of electrons and electromagnet lenses.

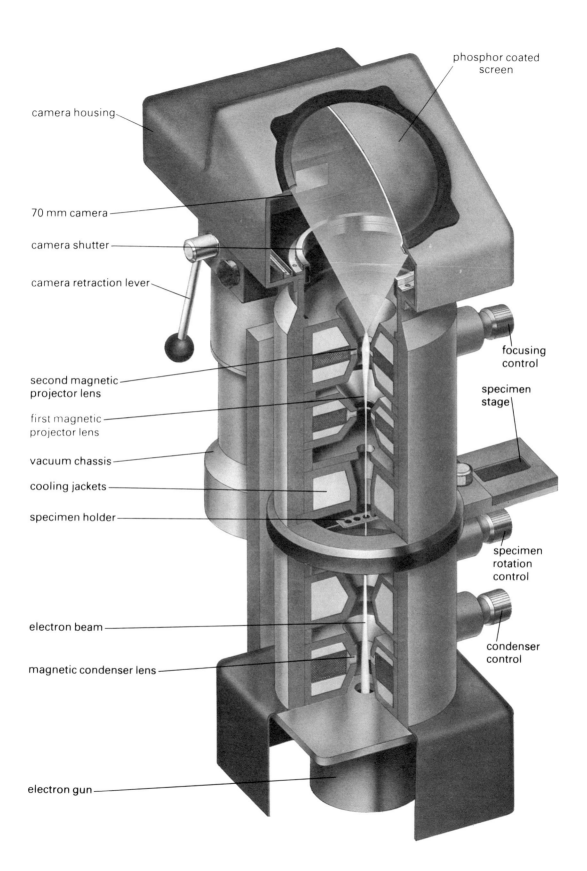

Opposite top: This power station at Runcorn, England, generates electricity for domestic, commercial and industrial use.
Opposite bottom: This huge diesel generator in a power station converts energy from oil into electricity.

specimens, higher voltages are necessary, making some types of electron microscopes very large and expensive.

The scanning electron microscope is used to study the surfaces of objects. The surface is scanned by a very fine beam of electrons which pass into it and scatter other electrons from its atoms. Some of these scattered electrons come out of the surface being examined, are picked up and passed to the screen of a cathode ray tube, like a TV picture tube. The picture obtained shows irregularities in the surface of the specimen as well as parts with heavy atoms or substances which do not conduct electricity.

Electron microscopes can magnify up to half a million times. They are used a great deal in medical research to study micro-organisms that cause disease, and to study the tiny structures of the body's cells.

# PRODUCING & DISTRIBUTING ELECTRICITY

Electricity is the world's most used form of power. It is used to run trains and factories, to light cities, to run all the electrical devices in modern homes and to operate radio and television services.

## Power stations

Below: This cross-sectional view through a modern electric AC generator clearly shows the internal workings. Both mechanically and electrically, this generator is very similar to an electric motor.

Electricity is generated at power stations which convert one form of energy into another. Burning *fossil fuels* such as coal produces heat which in turn heats water to

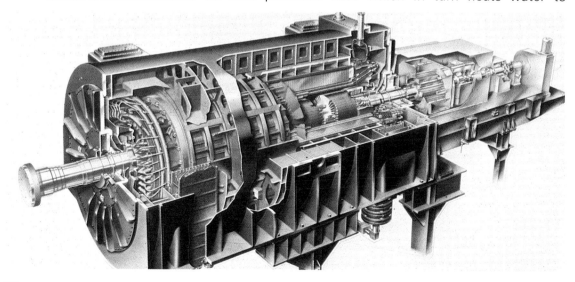

Right: Electric cables are very strong. This cross-section clearly shows the strands of copper inside a standard cable insulated with rubber.

Left: This is the national control room at the headquarters in London of the Central Electricity Generating Board (CEGB)

Below: Electricity is distributed through a system of cables and wires which is known as a power grid.
1. Power station.   2. Distribution/Transformer plant   3. Transmission pylon.   4. Transformer station   5. Transmission pylon.   6. Distribution system   7. Heavy industry   8. Transformer station   9. Light industry   10. Transformer station   11. Domestic usage.   12. Farms

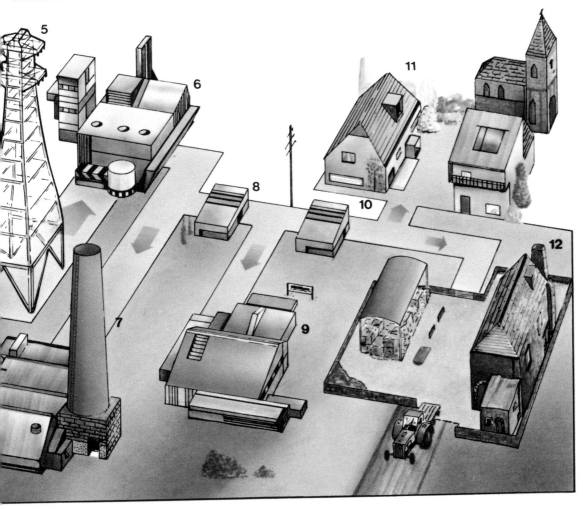

Above: The CEGB electrical sub-station at Wymondley, England, was designed to be as compact and space-saving as possible. The normal massive steel gantries have been replaced by a self-supporting lattice of aluminium in the form of a bridge. Electricity is transmitted to the sub-station by high-voltage power lines.

Opposite: This circuit-breaker assembly is being tested for operational accuracy on a special test bed. You can see the contact rods at the base.

produces steam, this in turn drives the turbine which drives the generator. In hydroelectric power stations falling water is directed through a turbine which generates electricity. Nuclear power plants and solar units likewise convert one form of energy into another. *Transformers* at power stations step up or increase the voltage of electricity from several thousand to many thousands of volts before it is sent through the power lines.

## The power grid

Electric power lines made from copper or aluminium are suspended from overhead pylons in open country and carried underground in towns and cities. In the United States, with its huge population, there are about 800,000 km of high-tension power cables, long enough to stretch to the moon and back. This system of cables is called a *grid*.

# Transformers

Because of the different equipment used to produce electricity, such as the size of the core, the number of coils of wire in a generator, and the speed at which it operates, electricity is produced in different voltages from different generators. On the other hand, the electric motors used vary from big, heavy-duty ones for industry to very small ones. It is necessary in some way to alter the voltages of the electricity according to how it is to be used. It is better to send the electricity out from the power-station in high voltage form so that little is lost in transmission. So it is often transmitted at about 30,000 volts, but it comes into our homes at about 240 volts, having been broken down at the local sub-station. The job of stepping electricity up or down in voltage is done by *transformers.*

A transformer usually consists of a soft iron core on

Below: Power stations, such as this one at Reading, England, generate the electricity used by factories, offices, transport systems and homes.

Below right: This enormous transformer in a power staion can increase the voltage of the electricity being produced.

which is wound two separate coils of insulated wire. One coil is called the *primary* coil and the other is the *secondary* coil.

If the voltage is to be increased, a *step-up transformer* is used. In a step-up transformer, the secondary coil has more turns than the primary coil. Electricity is fed into the primary coil and sets up magnetic lines of force. These are picked up on the secondary coils and because there are more of them the original voltage is amplified and let out through the secondary coil terminals.

The reverse happens when the voltage is to be decreased with a *step-down transformer.* In this case

These men are connecting a transformer to some overhead high-tension power lines. Transformers can increase or decrease the electrical voltage.

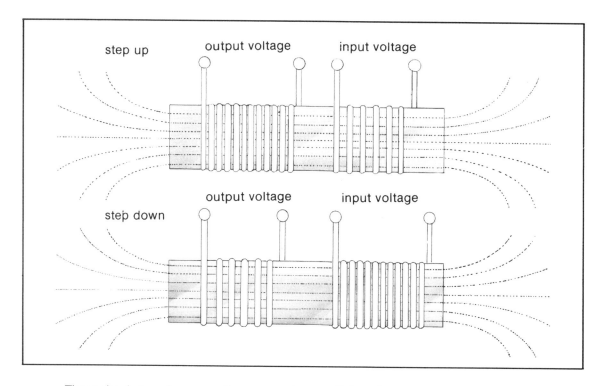

These simple transformers are used to alter the voltage of electricity. The transformer consists of a primary and a secondary coil of insulated wire wound around a soft iron core. A step-up transformer is used to increase voltage whereas a step-down transformer can decrease the voltage. the secondary coil has less turns than the primary. As electricity is fed into the primary coil it again sets up a magnetic field but because there are fewer turns in the secondary coil the voltage is decreased before it is let out through the secondary coil terminals.

An appliance such as a battery or mains cassette player is built with its circuits to take the output of the batteries, usually about 6 or 12 volts. If it were plugged straight into the mains power of 240 volts it would burn out, so it has a transformer built in to step down the mains voltage. This is why battery or mains radios and cassette players have unusual power plugs. It is partly to stop you using the wrong plug, but mostly because one of the pins is also a device to cut in the transformer before you switch on the power. A colour television set will have at least two transformers, because different parts of the set operate at different voltages.

Inside a transformer there are varnished coils, as in an electric motor. The iron core is usually not round but flat and is not one solid piece but is made up of thin plates or *laminations*. The laminations of iron will either be varnished or separated by thin layers of paper. This is to prevent the core itself absorbing current from the primary coil and so wasting energy and, also, to prevent the core from becoming too hot.

# INDEX

MAGNETISM AND ELECTRICITY 1-8
ELECTROMAGNETISM 9-12
ELECTRICITY 12-17
ELECTRIC GENERATORS AND MOTORS 18-24
CIRCUITS AND USES OF ELECTRICITY 25-32
PRODUCING & DISTRIBUTING ELECTRICITY 32-40

Page numbers in italics refer to a diagram on that page.
Bold type refers to a heading or sub-heading.

## A
Alternating current (AC) **20**, 21
Amber 12
Ammeter *15*
Ampere, Andre 9
Amplifiers 17
Armature 18, 22, 27-28
Armature winding 18-19, *18*

## B
Baird 17
Bar magnet **4-5**
 moving *15*
Battery 20, **21**, 22
 dry cell *21*
 wet cell 20
Bell, electric **27**, *27*, 28
Brushes *18*, 20

## C
Cables, electric 34, 36
Cambridge Stereoscan 30
Carbon arc lamps 27
Cathode ray tube 17
Circuit **25**
 parallel **25**, *25*
 series **25**, *25*
Circuit breaker 37
Columbus 2
Compass **1**, *1*, 2, *2*, 4, **7**, *7*, 8
 ship's **7**, *7*, 8
Conductors 12
Cooke 17

## D
Demagnetise 3
Diesel generator *33*
Direct current (DC) **20**, 21
Discharge lamps 26

## E
Earth's magnetism **6**, *6*, 7
Edison, Thomas 17
Electric
 battery 15
 bell **27**, *27*, 28
 cables 34, 36
 lights **26**, *26*, 27
 generator *15*, 16-20,
  *18, 19, 32*
 motor 12, **22**, 22-24,
  *22, 23, 24*
 telegraph *16*, 17
Electrical inventions 17

Electricity **12-17**
 production **15**, *15*, 16, 32-40
 static **12**, *12*, 14
Electrolysis 29
Electrolytic cells 21-22
Electromagnet 2, **11**, *11*, 12,
  18-19
Electromagnetic induction 16
Electromagnetism **9-12**, *9*
Electron microscope **30**, 30-32,
  *30, 31*
Electroplating **28**, *28*, 29
  *29*

## F
Faraday, Michael 9, 15-16
Ferromagnetic metals **3**, *3*
Filament 26
Fluorescent lamps 27
Franklin, Benjamin 14
Friction 14

## G
Generator **15**, 16, **18**,
  18-20, *18, 19, 32*
 diesel *33*
Gimbals 8
Gyrocompass **7**, 8, *8*
 Sperry CL11 *8*
Gyroscope 8

## H
Henry, Joseph 12
Horseshoe magnet **5**, *5*

## I
Induction **19**
 motor 23, *23*, 24
Insulated wire 10, *10*
Insulators 12, *13*
Iron 2
 soft 2
Iron ore 1

## L
Laminations 40
Lightning 12, 14
Linear induction motor 24
Linear motor 23
Lines of force 5
 magnetic 5, *11*
Lodestone **1**, 2

## M
Magnet 2, 3
 bar 2, 4, *4*, 5
 horseshoe 2, 5, *5*
Magnetic
 declination 7
 field 3, **4**, 5, *5*
 force 5
 poles **4**, 4-6,
  *4, 5, 6*
 variation 7
Magnetising **2**, 3
 metal **2**, 3
Magnetism 1
Magnetometer 7
Mariner's compass **7**, *7*, 8
Metal alloys 2
Metals 3
Morse, Samuel 17
Morse code 16

## N
Non-conductors 12
North Pole (earth) **6**, *6*, 7

## O
Oersted, Hans **9**, *9*
 experiment **9**, *9*, 10

## P
Parallel circuits **25**, *25*
Poles, magnetic 2, 3, *3*, **4**,
  *4, 5, 5*
Power grid 34, *35* **36**
Power stations **32**, 32-36,
  *33, 34, 35*, 38
Primary cells 22
Primary coil 39

## R
Radio communication 7
Rotor 22

## S
Secondary cells 22
Secondary coils 39
Series circuits **25**, *25*
Ship's compass **7**, *7*, 8
Slip rings *18*, 20
Solar winds 7
Solenoid **10**, *10*, 11
South Pole (earth) **6**, *6*, 7
Static electricity **12**, *12*, 14

Stator 23
Steel, hard 2
Storage cells 22
Sturgeon, William 11
Sub-station, electrical *36*
Sunspots 7
Synchronous motor 24

# T

Telegraph *16*
Television *17*, 17
Thales 12
Transformers 36, **38**, 38-40, *38, 39*
   step-down 39-40, *40*
   step-up 39-40, *40*
Turbine 18

# V

Volt 15
Volta, Alessandro 15
Voltaic cell 15
Voltaic pile 15

# W

Wheatstone 17
Wheatstone's telegraph *16*, 17
Wimshurst machine 14, *14*
Windings 18
Wire, insulated 10, *10*, 11